Thirteenth Interim Report of the Subcommittee on Acute Exposure Guideline Levels

Subcommittee on Acute Exposure Guideline Levels

Committee on Toxicology

Board on Environmental Studies and Toxicology

Division on Earth and Life Studies

NATIONAL RESEARCH COUNCIL
OF THE NATIONAL ACADEMIES

THE NATIONAL ACADEMIES PRESS
Washington, D.C.
www.nap.edu

THE NATIONAL ACADEMIES PRESS   500 Fifth Street, NW   Washington, DC 20001

NOTICE: The project that is the subject of this report was approved by the Governing Board of the National Research Council, whose members are drawn from the councils of the National Academy of Sciences, the National Academy of Engineering, and the Institute of Medicine. The members of the committee responsible for the report were chosen for their special competences and with regard for appropriate balance.

This project was supported by Contract Nos. DAMD 17-89-C-9086 and DAMD 17-99-C-9049 between the National Academy of Sciences and the U.S. Army. Any opinions, findings, conclusions, or recommendations expressed in this publication are those of the author(s) and do not necessarily reflect the view of the organizations or agencies that provided support for this project.

International Standard Book Number: 0-309-09707-X

Additional copies of this report are available from

The National Academies Press
500 Fifth Street, NW
Box 285
Washington, DC 20055

800-624-6242
202-334-3313 (in the Washington metropolitan area)
http://www.nap.edu

Copyright 2005 by the National Academy of Sciences. All rights reserved.

Printed in the United States of America

THE NATIONAL ACADEMIES
Advisers to the Nation on Science, Engineering, and Medicine

The **National Academy of Sciences** is a private, nonprofit, self-perpetuating society of distinguished scholars engaged in scientific and engineering research, dedicated to the furtherance of science and technology and to their use for the general welfare. Upon the authority of the charter granted to it by the Congress in 1863, the Academy has a mandate that requires it to advise the federal government on scientific and technical matters. Dr. Ralph J. Cicerone is president of the National Academy of Sciences.

The **National Academy of Engineering** was established in 1964, under the charter of the National Academy of Sciences, as a parallel organization of outstanding engineers. It is autonomous in its administration and in the selection of its members, sharing with the National Academy of Sciences the responsibility for advising the federal government. The National Academy of Engineering also sponsors engineering programs aimed at meeting national needs, encourages education and research, and recognizes the superior achievements of engineers. Dr. Wm. A. Wulf is president of the National Academy of Engineering.

The **Institute of Medicine** was established in 1970 by the National Academy of Sciences to secure the services of eminent members of appropriate professions in the examination of policy matters pertaining to the health of the public. The Institute acts under the responsibility given to the National Academy of Sciences by its congressional charter to be an adviser to the federal government and, upon its own initiative, to identify issues of medical care, research, and education. Dr. Harvey V. Fineberg is president of the Institute of Medicine.

The **National Research Council** was organized by the National Academy of Sciences in 1916 to associate the broad community of science and technology with the Academy's purposes of furthering knowledge and advising the federal government. Functioning in accordance with general policies determined by the Academy, the Council has become the principal operating agency of both the National Academy of Sciences and the National Academy of Engineering in providing services to the government, the public, and the scientific and engineering communities. The Council is administered jointly by both Academies and the Institute of Medicine. Dr. Bruce M. Alberts and Dr. Wm. A. Wulf are chair and vice chair, respectively, of the National Research Council.

www.national-academies.org

SUBCOMMITTEE ON ACUTE EXPOSURE GUIDELINE LEVELS

DONALD E. GARDNER *(Chair)*, Inhalation Toxicology Associates, Raleigh, NC
DANIEL KREWSKI *(past Chair)*, University of Ottawa, Ontario, Canada
EDWARD C. BISHOP, Parsons Corporation, Pasadena, CA
JAMES V. BRUCKNER, University of Georgia, Athens
RAKESH DIXIT, Merck and Company, Inc., West Point, PA
JOHN DOULL *(past member)*, University of Kansas Medical Center, Kansas City
JEFFREY W. FISHER, University of Georgia, Athens
DAVID W. GAYLOR *(past member)*, Gaylor and Associates, LLC, Eureka Springs, AR
KANNAN KRISHNAN *(past member)* University of Montreal, Quebec, Canada
DAVID P. KELLY, Dupont Company, Newark, DE
STEPHEN U. LESTER, Center for Health, Environment, and Justice, Falls Church, VA
JUDITH MACGREGOR, Toxicology Consulting Services, Arnold, MD
PATRICIA M. MCGINNIS *(past member)* Syracuse Research Corporation, Ft. Washington, PA
DAVID A. MACYS, Island County Health Department, Coupeville, WA
FRANZ OESCH, University of Mainz, Mainz, Germany
RICHARD B. SCHLESINGER, Pace University, New York, NY
CALVIN C. WILLHITE *(past member)*, California Department of Toxic Substances Control, Berkeley
FREDERIK A. DE WOLFF, Leiden University Medical Center, Leiden, Netherlands

Staff

KULBIR S. BAKSHI, Project Director
ALEXANDRA STUPPLE, Senior Editorial Assistant
AIDA C. NEEL, Program Associate

Sponsor

U.S. DEPARTMENT OF DEFENSE

COMMITTEE ON TOXICOLOGY

WILLIAM E. HALPERIN *(Chair)*, New Jersey Medical School, Newark
LAWRENCE S. BETTS, Eastern Virginia Medical School, Norfolk
EDWARD C. BISHOP, Parsons Corporation, Pasadena, CA
JAMES V. BRUCKNER, University of Georgia, Athens
GARY P. CARLSON, Purdue University, West Lafayette, IN
JANICE E. CHAMBERS, Mississippi State University, Mississippi State
MARION EHRICH, College of Veterinary Medicine, Blacksburg, VA
SIDNEY GREEN, Howard University, Washington, DC
MERYL KAROL, University of Pittsburgh, Pittsburgh, PA
JAMES MCDOUGAL, Wright State University School of Medicine, Dayton, OH
ROGER MCINTOSH, Science Applications International Corporation, Abingdon, MD
GERALD N. WOGAN, Massachusetts Institute of Technology, Cambridge

Staff

KULBIR S. BAKSHI, Program Director
EILEEN N. ABT, Senior Program Officer for Risk Analysis
SUSAN N. J. MARTEL, Senior Program Officer
ELLEN K. MANTUS, Senior Program Officer
ALEXANDRA STUPPLE, Senior Editorial Assistant
AIDA NEEL, Program Associate
TAMARA DAWSON, Program Assistant
SAM BARDLEY, Librarian

BOARD ON ENVIRONMENTAL STUDIES AND TOXICOLOGY[1]

Members

JONATHAN M. SAMET *(Chair)*, Johns Hopkins University, Baltimore, MD
RAMON ALVAREZ, Environmental Defense, Austin, TX
THOMAS BURKE, Johns Hopkins University, Baltimore, MD
JUDITH C. CHOW, Desert Research Institute, Reno, NV
COSTEL D. DENSON, University of Delaware, Newark
E. DONALD ELLIOTT, Wilkie, Farr & Gallagher, LLP, Washington, DC
CHRISTOPHER B. FIELD, Carnegie Institute of Washington, Stanford, CA
SHERRI W. GOODMAN, Center for Naval Analyses Corporation, Alexandria, VA
JUDITH A. GRAHAM, American Chemistry Council, Arlington, VA
DANIEL S. GREENBAUM, Health Effects Institute, Cambridge, MA
ROBERT HUGGETT, Michigan State University, East Lansing
BARRY L. JOHNSON, Emory University, Atlanta, GA
JAMES H. JOHNSON, Howard University, Washington, DC
JUDITH L. MEYER, University of Georgia, Athens
PATRICK Y. O'BRIEN, ChevronTexaco Energy Technology Company, Richmond, CA
DOROTHY E. PATTON, International Life Sciences Institute, Washington, DC
STEWARD T.A. PICKETT, Institute of Ecosystem Studies, Millbrook, NY
JOSEPH V. RODRICKS, ENVIRON Corporation, Arlington, VA
ARMISTEAD G. RUSSELL, Georgia Institute of Technology, Atlanta
MITCHELL J. SMALL, Carnegie Mellon University, Pittsburgh, PA
LISA SPEER, Natural Resources Defense Council, New York, NY
KIMBERLY M. THOMPSON, Harvard School of Public Health, Boston, MA
G. DAVID TILMAN, University of Minnesota, St. Paul
CHRIS G. WHIPPLE, ENVIRON Corporation, Emeryville, CA
LAUREN A. ZEISE, California Environmental Protection Agency, Oakland

Senior Staff

JAMES J. REISA, Director
DAVID J. POLICANSKY, Scholar
RAYMOND A. WASSEL, Senior Program Officer for Environmental Sciences and Engineering
KULBIR BAKSHI, Senior Program Officer for Toxicology
EILEEN N. ABT, Senior Program Officer for Risk Analysis
K. JOHN HOLMES, Senior Program Officer
SUSAN N.J. MARTEL, Senior Program Officer
SUZANNE VAN DRUNICK, Senior Program Officer
ELLEN K. MANTUS, Senior Program Officer
RUTH E. CROSSGROVE, Senior Editor

---

[1]This study was planned, overseen, and supported by the Board on Environmental Studies and Toxicology.

Preface

Extremely hazardous substances (EHSs)[1] can be released accidentally as a result of chemical spills, industrial explosions, fires, or accidents involving railroad cars or trucks transporting EHSs, or intentionally through terrorist activities. However, it is also feasible that these substances can also be released by improper storage and/or handling. Workers and residents in communities surrounding industrial facilities where EHSs are manufactured, used, or stored and in communities along the nation's railways and highways are potentially at risk of being exposed to airborne EHSs during accidental and intentional releases. Pursuant to the Superfund Amendments and Reauthorization Act of 1986, the U.S. Environmental Protection Agency (EPA) has identified approximately 400 EHSs on the basis of acute lethality data in rodents.

The National Advisory Committee (NAC) on Acute Exposure Guideline Levels for Hazardous Substances has developed acute exposure guideline levels (AEGLs) for approximately 120 EHSs to date. In 1998, EPA and the U.S. Department of Defense (DOD) requested that the National Research Council (NRC) independently review the AEGLs developed by the NAC. In response to that request, the NRC organized within its Committee on Toxicology the Subcommittee on Acute Exposure Guideline Levels. The NAC's Standing Operating Procedures for Developing AEGLs for Airborne Chemicals was reviewed by the subcommittee and published in May 2001. That report provides step-by-step guidance for the derivation of AEGLs for hazardous chemicals. In December 2000, the subcommittee's first report on specific chemicals, Acute Exposure Guideline Levels for Selected Airborne Chemicals, Volume 1, was published by the NRC; volumes 2, 3, and 4 in that series were published in 2002, 2003, and 2004 respectively.

The subcommittee meets two times each calendar year. At those meetings, the subcommittee hears presentations from the NAC staff and its contractor—the Oak Ridge National Laboratory—on draft AEGL documents. At some meetings, the subcommittee also hears presentations from NAC's collaborators from other countries, such as Germany. The subcommittee provides comments and recommendations on those documents to NAC in its interim reports, and the NAC uses those comments to make revisions. The revised reports are presented by the NAC to the subcommittee at subsequent meetings until the subcommittee concurs with the final draft documents. The revised reports are then published as appendices in the subcommittee's reports.

The present report is the subcommittee's thirteenth interim report. It summarizes the subcommittee's conclusions and recommendations for improving NAC's AEGL documents for 10 chemicals: 1, 4-dioxane; chloroform; carbon tetrachloride; sulfur dioxide; 1,-2 dichloroethylene; monochloroacetic acid; carbon monoxide; fluorine; methanol; and phenol.

This report has been reviewed in draft form by individuals chosen for their diverse perspectives and technical expertise, in accordance with procedures approved by the NRC's Report Review Committee. The purpose of this independent review is to provide candid and critical comments that will assist the institution in making its published report as sound as possible and to ensure that the report meets institutional standards for objectivity, evidence, and responsiveness to the study charge. The review comments and draft manuscript remain confidential to protect the integrity of the deliberative process. We wish to

---

[1] As defined pursuant to the Superfund Amendments and Reauthorization Act of 1986.

thank the following individuals for their review of this report: Deepak K. Bhalla (Wayne State University), David W. Gaylor (Gaylor and Associates, LLC), and Sam Kacew (University of Ottawa).

Although the reviewers listed above have provided many constructive comments and suggestions, they were not asked to endorse the conclusions or recommendations, nor did they see the final draft of the report before its release. The review of this report was overseen by: Sidney Green, Jr. (Howard University). Appointed by the NRC, he was responsible for making certain that an independent examination of this report was carried out in accordance with institutional procedures and that all review comments were carefully considered. Responsibility for the final content of this report rests entirely with the authoring committee and the institution.

The subcommittee gratefully acknowledges the valuable assistance provided by the following people: Susan Martel (Program Officer, NRC Committee on Toxicology), Ernest Falke, Iris Camacho, and Paul Tobin (all from EPA); Cheryl Bast and Robert Young (both from Oak Ridge National Laboratory); and Peter Griem of Germany. Aida Neel was the program associate and Alexandra Stupple was the editor. We are grateful to James J. Reisa, director of the Board on Environmental Studies and Toxicology, for his helpful guidance. The subcommittee particularly acknowledges Kulbir Bakshi, project director for the subcommittee, for bringing the report to completion. Finally, we would like to thank all members of the subcommittee for their expertise and dedicated effort throughout the development of this report.

Donald E. Gardner, *Chair,*
Subcommittee on Acute Exposure
Guideline Levels

William E. Halperin, *Chair*
Committee on Toxicology

Thirteenth Interim Report of the Subcommittee on Acute Exposure Guideline Levels

BACKGROUND

In 1991, the U.S. Environmental Protection Agency (EPA) and the Agency for Toxic Substances and Disease Registry (ATSDR) asked the National Research Council (NRC) to provide technical guidance for establishing community emergency exposure levels (CEELs) for extremely hazardous substances (EHSs) pursuant to the Superfund Amendments and Reauthorization Act of 1986. In response to that request, a subcommittee of the NRC Committee on Toxicology prepared a report titled *Guidelines for Developing Community Emergency Exposure Levels for Hazardous Substances* (NRC 1993). That report provides step-by-step guidance for the derivation of CEELs for EHSs.

In 1995, EPA, several other federal and state agencies, and several private organizations academia convened an advisory committee—the National Advisory Committee on Acute Exposure Guideline Levels (AEGLs) for Hazardous Substances (referred to as the NAC)—to develop, review, and approve AEGLs (similar to CEELs) for up to 400 EHSs. AEGLs developed by the NAC have a broad array of potential applications for federal, state, and local governments and for the private sector. AEGLs are needed for prevention and emergency-response planning for potential releases of EHSs, either from accidents or as a result of terrorist activities.

THE CHARGE TO THE SUBCOMMITTEE

The NRC convened the Subcommittee on Acute Exposure Guideline Levels to review the AEGL documents approved by the NAC. The subcommittee members were selected for their expertise in toxicology, pharmacology, medicine, industrial hygiene, biostatistics, risk assessment, and risk communication.

The charge to the subcommittee is to (1) review AEGLs developed by the NAC for scientific validity, completeness, and conformance to the NRC (1993) guidelines report, (2) identify priorities for research to fill data gaps, and (3) identify guidance issues that may require modification or further development based on the toxicological database for the chemicals reviewed.

This interim report presents the subcommittee's comments concerning the NAC's draft AEGL documents for 10 chemicals: 1, 4-dioxane; chloroform; carbon tetrachloride; sulfur dioxide; cis, trans 1,-2 dichloroethylene; monochloroacetic acid; carbon monoxide; fluorine; methanol; and phenol.

COMMENTS ON 1, 4-DIOXANE

At its February 21-23, 2005, meeting, the subcommittee reviewed the AEGL document on 1,4-dioxane. The presentation was made by Peter Griem of Clariant, Germany. The subcommittee recommends a number of revisions. The subcommittee will review the revised AEGLs draft at its future meeting.

General Comment

The subcommittee concludes that the document is well written. The major comments include using a physiologically based pharmacokinetic (PBPK) model for AEGL-2 and AEGL-3, determining whether CNS effects (AEGL 2) and delayed liver toxicity (AEGL-3) are the most relevant end points or whether CNS be used for both end points, and being consistent with the use of uncertainty factors (UFs).

Major Comments

The subcommittee noted that the ambient air EPA's proposed value of 0.061 $\mu g/m^3$ is based on cancer. Therefore, the proposed AEGL values do not protect against excess calculated cancer risks. Cancer risk was not discussed adequately. Some consideration on how to address this public health concern is warranted. Reword sentences on page 58, lines 1966-1967.

There are inconsistencies regarding the use of UFs:

Page 29, lines 1218-1219. In the section titled Interspecies Variability, revise to read "Taken together the interspecies variability for acute lethal effects is limited and an interspecies uncertainty factor of 3 is considered adequate."

Page 31, line 1236. In the section titled Intraspecies Variability, revise to read "Due to the lack of data, there was no basis for reducing the default intraspecies uncertainty factor."

Page 32, lines 1287-1288. "A total uncertainty factor of 3 was used because for local effects, the toxicokinetic differences do not vary considerably within and between species." The NAC needs to break this down to inter- and intra-, not combine them.

Page 34, lines 1359-1361. "An interspecies uncertainty factor of 1 was applied because metabolism in humans and rats is very similar, involving the same metabolic steps and intermediate metabolites (See Section 4.3.2)..." First, this should probably be Section 4.3.1, Pharmacokinetic Modelling. Second, the subcommittee did not see any documentation that the metabolic steps are similar. It is not that the subcommittee doubts it, only that there is no evidence presented in the toxicity summary document (TSD). There is a statement to that fact on page 29, lines 1216-1217, but Section 4.1 does not give data to support that statement.

The document needs to be consistent and in agreement with the Standard Operating Procedure (SOP) manual on the use of uncertainty factors (UFs).

AEGL-1

The proposed AEGL-1 value is appropriate for contact irritation. A short discussion is needed for the selection of the lowest exposure to humans that caused eye irritation. The idea is that the AEGL is a discomfort number; thus, the question is how high can the concentration go for irritation before other effects become important? Would 200 or 300 ppm work (Silverman et al. 1946), since this is a reported 8 h value based on a 15 min exposure? The subcommittee understands that 5,500 ppm is too high because other effects were observed at 1 min (Yant et al. 1930). The proposed value (17 ppm) appears to be low given that workplace exposures are 50 ppm or less. Table 2 should help on the selection. A UF of 3 may be appropriate, but not if the NAC uses 50 ppm as the starting place. Time scaling with no change in concentration seems appropriate for eye irritation.

NOTE: The Young study is the proper basis. While the subcommittee agrees with the use of the UF of 3, it does not agree with the statement that toxicokinetic differences do not vary considerably within or between species. Also, the subcommittee has stated that when irritation is the effect, the AEGL values will not vary across time.

AEGL-2

The use of administered dose (and scaling) is far inferior to the use of internal measures of dose (for relating internal effects to dose), and where appropriate, internal measures of dose should be used. To that end, the PBPK model of Reitz can be used to obtain internal measures of dose for 1,2-dioxane, which are better related to effect. The PBPK model of Reitz was reviewed by the authors, but no indication of its potential use was indicated. There is now a 20-y history of using PBPK models. Pharmacokinetics (PK) is considered necessary for drug usage and development. Evaluate CNS versus delayed liver (or kidney) toxicity as critical end points more carefully.

The primary studies cited are those of Drew (1978) and Goldberg (1964). However, in the discussion of the derivation, Yant (1930) and other older studies are cited. Focus on Goldberg (1964) study and use the others as support, not the other way around. The liver toxicity is subclinical and transient and is not a good end point. Stick with the CNS depression effect.

For solvents that cause CNS effects, the AEGL-2 and AEGL-3 values should remain relatively constant over a time exceeding 1 h due to blood equilibrium concentrations. The time to equilibrium depends upon the blood solubility of the compound (the more lipophilic, the quicker the equilibrium). Haber's rule ($C^n \times t$) does not apply. At some point, the compound reaches saturation. This must be determined with PK modeling to estimate blood concentration. What happens if there is no PK model? Observations of the effects at the concentration of concern are needed. Short-term (less than equilibrium time) duration to the longer duration cannot be extrapolated.

Solvents

For CNS depression where the compound is direct acting (not a metabolite[s]) small animals receive a larger dose than humans because of their more rapid respiration. Therefore, the interspecies UF should be 2 or less. However, the intraspecies UF should be greater than 3 (others stated these effects do not vary more than 2-3?).

AEGL-3

In setting AEGLs, risk assessors are more concerned with acute symptoms (CNS) and delayed effects (liver) of acute exposure than with possible carcinogenicity after long-term high-dose exposure. It is clear that death is not an acute effect of dioxane but rather a delayed effect of acute exposure. In addition to the fading of the odor threshold, this is an extra risk of exposure to dioxane.

The key study is appropriate. The AEGL value of 950 ppm is less than half the immediately dangerous to life and health concentration (IDLH). The subcommittee expects them to be more similar with the AEGL-3.

The subcommittee does not agree with the explanation for the UF of 10. The NAC has placed a lot of credibility on the Yant study (1930), but this study is not described in any detail. In fact, page 4, lines 394-395, states "The specifications of the chamber, the purity of the dioxane and the methods of generating and measuring the dioxane atmospheres were not reported." The subcommittee recommends referring to other studies for comparisons and developing a better argument for a UF of 10.

Minor Points

Page vii, lines 165-168, and page vi, lines 162-164. Insert dioxane concentrations used in various in vitro assays.

Page vii, lines 214-215. List specific liver enzymes.

Page vii, lines 224-241. Provide the rationale for the intraspecies UF of 10. Is variability in human metabolism greater than 10? Discuss the elimination rate to derive an empirical UF. This can be part of the PBPK approach.

Page viii, lines 234-236. While correct, the NAC should refer to the SOP: 2.2.2.3.2 Highest Exposure Level That Does Not Cause Lethality—Estimate Lethality Threshold—One-third of the LC_{50}.

Page 2, line 325. It seems obvious that dermal exposure does occur. The subcommittee recommend deleting the sentence: "No fatalities have been reported after oral or dermal contact with 1,4-dioxane." This might be true, but there is little doubt dermal exposure contributes, and this sentence tends to discount it.

Page 2, line 334. "The exposures probably involved inhalation and dermal contact," in discussing 5 deaths out of 16 exposed.

Page 2, lines 344-352. Describes death after exposure to 208-650 ppm and dermal exposure.

The subcommittee found the Johnstone (1959) reported concentrations (for death) to be suspect relative to other data. All other data sets seem appropriate and consistent with each other.

Line 125. The odor does not disappear, but the sensitivity of the olfactory system does. This implies an increase in overall risk as the natural warning system becomes impaired at longer exposure time.

Lines 133-139. The delayed response of the liver and kidneys apparently forms a greater risk than initial CNS depression.

Lines 207 and 241. What is the reason for the high intraspecies UF of 10?

Line 231. The AEGL-3 values are based on an article from 1959. This was published before the general use of gas chromatography. How reliable is the 4-h LC_{50} of 14,300 ppm in rats?

Line 347. See comment on line 231.

Line 396. Add comments that this study design (scientist acting as volunteer for own study) is no longer acceptable.

Lines 418-428. There is a 50-100 times difference in odor detection and threshold between the two cited studies. What could the explanation be? The Wirth and Klimmer (1936) results (lines 429-431) agree well with the Hellman and Small (1974) study.

Lines 432-445. The increased aspartame amino transferase/alanine amino transferase (ALAT/ASAT) activity can apparently not be attributed to dioxane exposure but rather to alcohol abuse. This should be stated more clearly because it now seems as if it is the combination of dioxane and ethanol (EtOH) that caused liver function impairment.

Line 460. The fact that 14 of 42 people died from cancer does not mean the dioxane is a human carcinogen. About 15 out of 42 people are likely to die from cancer eventually anyway. What was the age of death of these four patients?

Line 471. See comment on line 125. Apparent disappearance ("fading") of odor increases the risk of dioxane exposure.

Lines 496-498. What were the variation and range of the mentioned odor detection and recognition thresholds?

Line 577. Add a line on the ethical unacceptability of these experiments.

Line 790. How many is "about 6 out of 8" rats? Delete "about."

Line 887. Typo: genitoxicity → genotoxicity.

Lines 933-940. What is the biological plausibility of the appearance of nasal cancer in the high-dose group? From the genotoxicity studies, dioxane does not seem to be a (genotoxic) carcinogen (lines 482); it might only have some epigenetic effects at very high doses.

Line 992. Apart from a moderate CNS depression, other CNS effects have not been reported.

Lines 1113-1114 and 1132-1152. See comments on lines 933-940. In setting AEGLs, risk assessors are more concerned with acute symptoms and delayed effects of acute exposure than with possible carcinogenicity after long-term high-dose exposure.

Lines 1147-1152. A possibility is that one or more dioxane metabolites are more reactive than the parent compound.

Lines 1187-1194. The section on areas under the curve is not clear.

Line 1303. Vertigo impairs the ability to escape.

Line 1320. Again, delete "about."

The Executive Summary should contain information about the use and quantity of 1,4 dioxane produced.

The proper abbreviation for milliliters is mL, not ml.

COMMENTS ON CHLOROFORM

At its February 21-23, 2005, meeting, the subcommittee reviewed the AEGL document on chloroform. The document was presented by Robert Young of Oak Ridge National Laboratory. The subcommittee recommends a number of revisions. The subcommittee will review the revised AEGLs draft at its next meeting.

General Comments

The subcommittee concurs with AEGL-1 and AEGL-2 values; AEGL-3 values were judged to be somewhat overly conservative by some subcommittee members. One subcommittee member suggested using a PBPK model for AEGL-3.

Specific Comments

AEGL-1

The statement "Not recommended due to properties of the chemical; AEGL-1 effects unlikely to occur in the absence of notable toxicity," needs additional explanation to prevent users from assuming effects go from none to AEGL-2 rapidly and with no warning. This is not the case for chloroform as it has been for previous chemicals with steep dose-response curves. Explain that there are no appropriate AEGL-1 effects (the first effects noted are AEGL-2 effects) or supply some context for the lack of AEGL-1 values.

It is not possible to determine AEGL-1, because of the lack of data needed to set AEGL values based on irritation.

AEGL-2

AEGL-2 is conservative, but the subcommittee agrees with the use of fetotoxicity data. It would be better to select a nonrecurring dose—that is, a single dose.

AEGL-3

The AEGL-3 values appear to be somewhat conservative in light of the anesthesia data collected by Whitaker and Jones. The derivation using the animal data would normally be appropriate. However, this is a rare case where there is controlled human data at a concentration around the AEGL-3. A heterogeneous group of 1,500 human subjects was subjected to concentrations up to 22,500 ppm for up to 2 h without death (with the exception of one questionable case). Since this group of patients was already somewhat compromised (needing surgery), the subcommittee believes that this data should be used with an appropriate modifying factor but no intraspecies UF. The subcommittee believes that a 1-h value of about 7,000 ppm might be more appropriate given this unique human data base.

The AEGL-1 and AEGL-2 values are appropriate based on the data.

A chloroform PBPK model developed by R.A. Corley (2000) should be used for the AEGL-3 determination per the guidance document that is under development.

Additional References

Corley, R.A., S.M. Gordon, and L.A. Wallace. 2000. Physiologically based pharmacokinetic modeling of the temperature-dependent dermal absorption of chloroform by humans following bath water exposures. Toxicol. Sci. 53:13-23.

Delic, J.I., P.D. Lilly, A.J. MacDonald, and G.D. Loizou. 2000. The utility of PBPK in the safety assessment of chloroform and carbon tetrachloride. Regul. Toxicol. Pharmacol. 32(2):144-155.

Tan, Y.M., B.E. Butterworth, M.L. Gargas, and R.B. Conolly. 2003. Biologically motivated computational modeling of chloroform cytolethality and regenerative cellular proliferation. Toxicol. Sci. 75:192-200.

Other Comments

Page 8, lines 1, 12. Odor information repeats.

Page 9. Human anesthesia data: There is another clinical anesthesia paper that may have higher concentrations (even higher than the 22,500 ppm noted above). The NAC should review the following: Waters, R.M. 1951. Chloroform: A Study After 100 Years. Madison, Wisconsin: University of Wisconsin Press.

Page 13, Table 2. Add: 389 ppm, 30 min, no complaints (according to Lehman and assistants). Lehman, K.B. and F. Flury, eds. 1943. Pp. 138-145 in Toxicology and Hygiene of Industrial Solvents (English Translation). Baltimore, Maryland: The Williams and Watkins Co.

Page 14, line 18. Add: Nominal chloroform exposure concentrations were reported, and chamber concentrations were not analytically measured.

Page 18, lines 4-6. Haskell Laboratories should be Haskell Laboratory.

Page 30, lines 12-13. Were the kidney lesions only in male rats?

Page 35, line 20. Provide some explanation for this conclusion.

Section 6 and 7. The explanations accompanying the derivations of AEGL-2 and AEGL-3 are confusing, and it was unclear what values were chosen as the basis for these. The NAC should provide a clearer discussion.

Table 14. The ERPG-3 is higher than the AEGL-3. Does this appear consistent given the dose-response information available for chloroform?

COMMENTS ON CARBON TETRACHLORIDE

At its February 21-23, 2005, meeting, the subcommittee reviewed the AEGL document on carbon tetrachloride (CCl_4). The document was presented by Robert Young of Oak Ridge National Laboratory. The subcommittee recommends a number of revisions. The revised document will be reviewed by the subcommittee at its next meeting.

General Comment

The NAC should check to see whether there are any reports that CCl_4-protein adducts serve as neoantigens, since halothane adducts can in the halothane-induced hepatotoxicity.

Specific Comments

Page 2, line 6, and page 7, line 5. Few non-health professionals will know what "anthelmintic" means. Perhaps "kills worms" could be included in parentheses.

Page 3, lines 12-15. Why would the AEGL values not be appropriate if the potential for dermal absorption exists? CCl_4 is absorbed through the skin, although the extent of its uptake is minor when compared to inhalation (Stewart and Dodd 1964). These investigators assessed percutaneous absorption of CCl_4 and other halocarbons in human subjects under extreme conditions (immersion of a thumb in undiluted CCl_4 for 30 min). Systemic uptake of CCl_4 vapor through the intact skin would be much lower.

It is also stated here that "the possibility exists for long-term hepatotoxic effects possibly requiring the need for antioxidant therapy." What is the source of concern about long-term/ongoing hepatotoxicity from an acute exposure? CCl_4 that is absorbed from the lungs reaches the liver, undergoes metabolic activation, and damages the liver within a matter of a few minutes. This is adequate time for ingested CCl_4 to do the same (Rao and Reckangel 1968, 1969; Sanzgiri et al. 1997). Thus, antioxidant therapy is not beneficial in acute poisonings unless it is instituted very quickly.

Page 7, lines 5-6. Should "chloroform" be "carbon tetrachloride"?

Page 8, line 4. Umiker and Pearce (1955) are not included in the References.

Page 8, lines 10-26. The case described here is a classic example of potentiation of CCl_4 acute toxicity by an alcohol. The heavy drinker died of nephrotoxicity, while the two other individuals who were subjected to the same exposure conditions experienced only mild CNS effects. The phenomenon of alcohol potentiation should be pointed out.

Page 8, lines 31-34. It should also be stated that hepatotoxicity is commonly manifest in humans. The severity of liver injury, however, is frequently less than kidney injury.

Guild et al. (1959) in the text is cited as Guild et al. (1958) in the References.

Page 9, lines 26-35: Point out that the workers studied by Smyth et al. (1936) and Elkins (1942) were exposed subchronically, not acutely, to CCl_4. Peak vapor concentrations were likely much higher than the range of 20-85 ppm proposed by Elkins. These exposure concentrations are unreliable, and the daily exposure durations are unknown.

Page 9, lines 36-42. The reconstruction of the exposure scenario by Norwood et al. (1950) is very doubtful. CCl_4 has a saturated vapor pressure of 15.2 kPa (114 mm Hg). This corresponds to a concentration of 150,000 ppm. A dilution factor of 600 would be required to attain 250 ppm. Assuming the distance from the floor to the breathing zone is 6 feet, hygienists commonly use a dilution factor of 10. Thus, it is likely that the victim inhaled much more than 250 ppm.

Page 10, lines 1-4. Information is needed on the route, duration, and intensity of CCl_4 exposure(s) required to produce neurological damage.

Page 14, line 21. Should Stewart et al. be Adams et al.?

Page 16, line 31. Gehring (1987) is not included in the References.

Page 18, lines 3-5. This sentence is incomplete.

Page 18, lines 19-22. It is important to state that the experiment of Adams et al. (1952) that is described here involves acute inhalation exposure.

Page 20, Table 8. Cornish and Block monitored serum glutamic oxalo acetic transaminase (SGOT), whereas David et al. (1981) monitored serum glutamic pyruvic transaminase (SGPT). Include both in the heading on line 4. What does "U/L" stand for?

Page 20, lines 19-35. It is not necessary to include this account of the investigation of Van Stee et al. (1982). Assessment of their more sensitive measure of liver injury failed, and excessive variability of histopathological changes precluded the authors' ability to distinguish the effects of one exposure regimen from another.

Page 20, line 27. This sentence is incomplete.

Page 21, lines 23-31. The account of the study of Sakata et al. (1987) should also be deleted. Their methodology as described and their reported findings are not credible. Inhalation of 180 ppm for 15 min will not cause liver damage in rats unless they are pretreated with a cytochrome P450 inducer.

Page 21, lines 32 and 43. The introductory sentences state that an objective of each study was to examine the effect of the route of exposure on CCl_4 acute toxicity. No information is provided in the paragraph on this topic. Other study objectives were the delineation of effects of route and pattern of exposure of CCl_4 on its pharmacokinetics and metabolism. It would be best to merely state than Sanzgiri et al. (1995) and Wang et al. (1995) assessed the influence of inhaled CCl_4 on acute hepatotoxicity in rats.

Page 21, line 42. All abbreviations (for example, G6Pase) should be spelled out the initial time they appear in the text.

Wang et al. (1995) is not included in the References.

The NAC should be consistent with the nomenclature of the serum enzymes. Alanine-aminotransferase and glutamic-pyruvic transaminase are synonymous, as are aspartate-aminotransferase and glutamic-oxaloacetic transaminase.

Page 22, line 9. Should Section 3.2.4 be 3.1.4?

Page 22, line 11. Define ECt_{50}.

Page 23, lines 1-5 and Table 9. This is a summary table of nonlethal effects of CCl_4 in animals. It summarizes many studies, but it is not clear which studies are most relevant for deriving AEGLs. The accompanying text states "although data pertaining to acute exposures [are] the primary focus, longer-term exposures with observations at 24 hours or less are included as well as longer-term exposures that may provide useful perspective in assessing the effects of inhalation exposure to carbon tetrachloride." The table should be reorganized so that the acute studies that are germane to the derivation of AEGL values are identified separately from the (1) longer-term studies with observations at 24 h or less that may be relevant to the derivation of AEGLs and (2) from other longer-term studies that may provide support or other perspectives on the results from acute studies.

Page 23, lines 8-9. Do the available data allow any judgement(s) about relative susceptibilities of different species to lethality or to hepatic or renal toxicity?

Page 26, line 21. McGregor and Lang (1996) is not included in the References.

Page 26, lines 23-24. Loveday et al. (1990) in the text is listed as Loveday et al. (1991) in the References.

There is a more recent study by Nagano et al. (1998) that specifies everything that the authors of the current document rightfully criticize as missing in the Costa (1963) study. The Nagano study specifies the age, sex, and strain of the animals used as well as the concentrations and exposures. Moreover, the groups of animals were sufficiently large (n = 50), and both sexes of two species (mice and rats) were used. Besides liver tumors in both sexes of both species, adrenal pheochromocytomas were observed in the mice.

Page 27, lines 3-5. It is also important to point out here that it is inappropriate to rely on the aforementioned investigations because they involved chronic exposures.

Page 27, lines 15-17. It should be recognized that CCl$_4$-induced nephrotoxicity is much more frequent and pronounced in humans than in rodents.

Page 27, line 43. Lehmann and Schmidt-Kehl (1936) is not included in the References.

Page 28, lines 12-18. The NAC has cited a general conclusion of Paustenbach et al. (1988) that their PBPK model predicted that rats, monkeys, and humans metabolize and eliminate CCl$_4$ in a similar manner at inhaled concentrations up to 100 ppm. Careful examination of Paustenbach et al.'s blood and fat CCl$_4$ time-course simulations (Figures 10 and 11) reveal that this is not the case. Their model predicts that fat and venous blood CCl$_4$ concentrations are substantially higher in rats than in humans inhaling 5 ppm. PBPK modeling by Delic et al. (2000) revealed much greater metabolism of CCl$_4$ by rats and mice than by humans.

El-Masri et al. (1996), Smith et al. (1998), Thrall et al. (2000), and Fisher et al. (2003) have also published PBPK models for CCl$_4$ in animals. Utilization of one of these models for time scaling in derivation of AEGL-1, -2, and -3 values could be considered. It would be preferable, of course, to have a human PBPK model. Paustenbach et al.'s rat model did accurately predict postexposure exhaled breath concentrations of CCl$_4$ by human subjects, when the model parameters were scaled up to humans.

Page 29, line 25. Cohen 1957 is not included in the References.

Page 29, line 33. The title of the subheading should be "Species and Strain Variability."

Page 29, line 38. Andersen 1981 and Reitz et al. 1982 are not included in the References.

Page 29, lines 38-40. It should also be pointed out here that rats will receive a greater internal dose than humans upon equivalent inhalation exposures because of rats' higher respiratory rate, cardiac output, and blood:air partition coefficient. More importantly, rats metabolize substantially more CCl$_4$ than do humans because of rats' higher hepatic blood flow and higher cytochrome P4502E1 (CYP2E1) activity. Therefore, rats absorb more CCl$_4$, metabolically activate more, and typically experience more-pronounced liver injury than do humans.

A patient described by Folland et al. (1976) exhibited only a modest, transient increase in serum transaminase activity but experienced renal failure. This individual was thought to have inhaled CCl$_4$ on only 1 d, but had been preexposed to isopropanol, which induces CYP2E1 and thereby markedly potentiates acute CCl$_4$ cytotoxicity.

Page 29, lines 43-47. It should be understood that Stewart et al. (1961) only saw slight changes in serum iron and urinary urobilinogen in one or two of six subjects. These changes are not indicative of acute hepatocellular injury. Serum transaminase activities were unaffected. In contrast, Cornish and Block (1960) and David et al. (1981) observed modest increases in SGOT and SGPT activities in rats inhaling 250 ppm for 250 and 70 min, respectively. These were LOAELs, whereas the 70-min, 49-ppm, and the 180-min, 11-ppm human exposure regimens of Stewart et al. (1961) were NOAELs for elevation of serum transminases. For rats, 50 and 100 ppm (4 h) were NOAELs (Cornish and Block 1960). In summary, it is very difficult to draw conclusions about the relative susceptibility of rats and humans to CCl$_4$-induced acute hepatotoxicity from the aforementioned studies because of differences in the durations of the exposure regimens.

Thirteenth Interim Report of the Subcommittee on Acute Exposure Guideline Levels

Page 30, lines 6-13. In light of the foregoing, Table 11 should be deleted.

Page 30, lines 15-21. Might children be at greater risk of acute hepatorenal toxicity? Neonates and young infants (up to 90-d-old) generally have relatively low hepatic microsomal CYP2E1 levels (Johnsrud et al. 2003) and should be less susceptible than adults. Young children, on the other hand, have a higher capacity than older children, adolescents and adults to metabolize a number of drugs (Ginsberg et al. 2002). Young children might therefore be more susceptible than adults to CCl_4 but should be adequately protected by the intraspecies UF of 10.

Page 30, line 25. Folland et al. (1976) described potentiation of CCl_4 hepatorenal toxicity in humans not laboratory animals.

Page 31, line 5. It should be pointed out here that Cornish et al. (1967) pretreated their rats with alcohols. A period of 12-24 h is usually required for alcohols to induce P450s. In contrast, concurrent exposure to alcohols and CCl_4 may reduce CCl_4 metabolic activation and toxicity via competitive metabolic inhibition.

Page 31, lines 16-17, and page 32, line 6. The exposure concentrations at which Tomenson et al. (1995) observed were ≥ 4 ppm not ≤ 4 ppm. Their report of serum enzyme alterations at ≥ 4 ppm is not consistent with the findings of other investigators.

Page 31, lines 20-23. As previously mentioned, Stewart et al. (1961) did not see changes in SGOT.

Page 31, line 30. The introductory sentence indicates that studies that show "no effects" are consistent with the definition of AEGL-1 effects. This is not correct. Adverse health effects that are consistent with AEGL-1 effects are defined on page 1 as those that exposed individuals might "experience notable discomfort, irritation, or certain asymptomatic, non-sensory effects."

Page 32, lines 12-22. Some more details about the AEGL-1 derivation should be provided here in the text (e.g., UF and its basis, toxicity end point). The 4- and 8-h AEGL-1 values should be rounded off.

An interspecies UF of 10 is not warranted for calculation of AEGL-1 values. There are numerous reports in the anesthesiology literature that demonstrate that variability in the response of the CNS to volatile organic anesthetics in humans (including infants and the elderly) is quite modest (Gregory et al. 1969; de Jong et al. 1975; Stevens et al. 1975). An intraspecies UF of 3 is thus appropriate for CNS depression. Hepatorenal toxicity is very unlikely at CCl_4 exposure concentrations in the range proposed for the AEGL-1 values, even in an individual with induced/heightened CYP2E1 activity. A heavy drinker who developed hepatorenal toxicity was exposed to CCl_4 at an estimated 250 ppm (Norwood et al. 1950). As related previously, the actual inhaled concentration was probably much higher. Cornish et al. (1967) had to expose rats to 1,000 ppm for 2-2.5 h in order to demonstrate potentiation of hepatotoxicity by alcohols. A total UF of 3 would be appropriate for deriving AEGL-1 values.

Alternatively, Davis's (1934) NOAEL for CNS effects and renal injury of 76 ppm for 4 h could be utilized as a departure point for derivation of AEGL-1 values. Extrapolation from an intermediate duration exposure (4 h) to longer and shorter times would be preferable to extrapolation from 30 min to the longer exposure times. The scientific validity of using $C^n \times t = K$ under such circumstances for volatile organic chemicals (VOCs), particularly when data are not available to derive n for the appropriate end point, is highly questionably. Concern is allayed to some extent by the time course of CCl_4 in the blood of rats during inhalation of the chemical (Sanzgiri et al. 1995). Blood concentrations gradually increase rather than attain near steady-state, because of CCl_4's slow metabolism and suicide enzyme inhibition (Manno et al. 1992; Fisher et al. 2004). Nevertheless, the 5-fold decrease from the 10-min to the 8-h AEGL-1 is excessive.

Page 32, lines 23-24. It is stated here that the AEGL-1 values from alternate data sets are provided in Appendix B. Appendix B presents other data.

Page 33, lines 5-21. It should be related here that the applicability of rodent data to humans is limited by the difference in their major target organs. In rodents, the liver is primarily affected by CCl_4. In CCl_4-poisoned humans, liver injury is often relatively minor, whereas kidney damage predominates.

Page 33, lines 9-11 and 21. The effects described in lines 9-11 are said to be consistent with AEGL-2 effects, yet it is concluded in line 21 that "available animal data do not suggest a severity that is consistent with AEGL-2 effects."

Page 33, lines 23-37. CNS effects also serve as the toxic end point for the AEGL-2 values, but potentiation of hepatotoxicity is again used as the rationale for the intraspecies UF of 10. As previously described in comments on the AEGL-1 derivation, an intraspecies UF of 3 should be utilized rather than 10. The lowest reliable LOAEL (317 ppm for 30 min) should be used as the basis for AEGL-2. All three subjects who inhaled 317 ppm for 30 min apparently experienced headache, nausea, and vomiting, whereas only one individual found 1,191 ppm to be intolerable after 9 min. Elkins (1942) reported nausea and vomiting in workers inhaling undocumented vapor concentrations of 20-85 ppm, but these cases involved repeated exposures. The reasoning for the rejection and selection of particular studies and exposures should be clearly stated in the text.

Findings in human studies should serve as a reality check for the proposed AEGL-1 and AEGL-2 values. Stewart et al. (1961) found, for example, that inhalation of 49 ppm for 70 minutes was without ill effect in human subjects and inhalation of 11 ppm for 3 h produced no adverse effects. Davis (1934) reported no CNS depression nor kidney injury in humans inhaling 76 ppm for 4 h, but one worker exposed to 200 ppm for 8 h exhibited compromised kidney function.

Time scaling using $C^n \times t = k$ produces overly conservative AEGL-2 values for the longer exposure periods. As related on page 19, David et al. (1981) found that short exposures to high vapor concentrations produced larger rises in SGPT in rats than did longer exposures to lower concentrations despite equivalent ppm-h. Kim et al. (1990) found the highest correlation to be between liver damage and peak blood concentration (10-20 min postingestion), rather than area under the blood concentration versus time curve (AUC^∞_0) in CCl_4-dosed rats.

Page 34, lines 10-13. It should be noted here that the fatality involved a heavy drinker.

Page 34, line 29. There is inconsistency in discussions of which studies were used to derive the AEGL-3 values. The text here, in the Summary (page 2, lines 41-42) and in the Summary Table in Appendix D all identify two studies, Adams (1952) and Dow Chemical (1986), as the basis for deriving the

AEGL-3 values. Appendix A, however, lists three studies, the two studies mentioned above and a 1946 Union Carbide study. This third study is also included in Appendix B, which provides the details of the time scaling calculations used to derive the AEGL-3 values. Which are the key studies? Please clarify.

Page 34, lines 33 and 35. Appendix C should be B.

Page 34, lines 40-42. It is true that rats eliminate CCl_4 more rapidly than humans because of more rapid exhalation and metabolism. It is important to recognize, however, that with equivalent inhalation exposures, rats achieve much higher target organ doses because of their relatively rapid respiration, cardiac output, and tissue (for example, liver) blood flow rates, as well as higher blood:air partition coefficients (Gargas et al. 1989). Rats' more rapid CCl_4 metabolism is a two-edged sword in that CCl_4 is more rapidly eliminated by rats than by humans, but greater metabolism equates to increased metabolic activation and susceptibility to hepatotoxicity in rats. Delic et al. (2000) concluded from their work with a PBPK CCl_4 model for rats and humans that rats metabolize much more CCl_4 than humans and accordingly are more susceptible to CCl_4 (metabolite)-induced cytotoxicity. It is generally accepted that CCl_4 and other lipophilic volatile organics partition into neuronal membranes and cause CNS depression by the disruption of nerve impulses. Data from studies of aquatic organisms indicate that the critical brain concentration of a halocarbon required to produce a given level of narcosis is relatively constant across species (McCarty et al. 1991). The possibility remains, nevertheless, that molecular mechanisms of narcosis could vary among different species resulting in heretofore unrecognized interspecies pharmacodynamic differences in sensitivity to halocarbon-induced CNS depression. A detailed discussion of the foregoing is recommended in the text.

Page 35, lines 38-40. The discussion of potential cancer risks of an acute inhalation exposure to CCl_4 is incomplete and inaccurate. As previously pointed out, Nagano et al. (1998) have published the results of an inhalation carcinogenicity study with clear-cut findings from which cancer risks can be calculated. The SOP manual (page 111) calls for the use of a three-risk range, from 10^{-4} to 10^{-6}. Estimation of just one risk value is included in Appendix C.

The NAC should give a more complete explanation of why the AEGL values for CCl_4 are not based upon cancer risk estimates. Make it clear that AEGL values based on unit cancer-risk values were higher than when they were based upon noncancer end points. This explanation and reasoning should be included here in the text, in the Summary, and in Appendix C.

The International Agency for Research on Cancer (IARC) and the National Toxicology Program (NTP) carcinogenicity classifications should be included in lines 11 and 12 of page 27. It should also be noted here or elsewhere that ATSDR (2004) has published an updated *Toxicological Profile for CCl_4*.

Page 36, lines 23-25. Shouldn't the Amercian Conference of Governmental Industrial Hygienists threshold limit value-time weighted average (ACGIH TLV-TWA) and short-term exposure limit (STEL) be included in the 8-h column?

Page 52, Appendix C, line 3. Is the risk level associated with a CCl_4 vapor concentration of 7E-2 $\mu g/m^3$ 1 in 100,000 or 1 in 1,000,000? As noted above, three risk values should be calculated.

Additional References

ATSDR (Agency for Toxic Substances and Disease Registry). 2004. Toxicological Profile for Carbon Tetrachloride. Agency for Toxic Substances and Disease Registry, Atlanta, GA.

de Jong, R.H., and E.I. Eger. 1975. AD_{50} and AD_{95} values of common inhalation anesthetics in man. Anesthesiology 42:384-389.

Delic, J.I., P.D. Lilly, A.J. MacDonald, and G.D. Loizou. 2000. The utility of PBPK in the safety assessment of chloroform and carbon tetrachloride. Regul. Toxicol. Pharmacol. 32:144-155.

El-Masri, H.A., R.S. Thomas, G.R. Sabados, J.K. Phillips, M.E. Andersen, H.M. Mehendale, and R.S.H. Yang. 1996. Physiologically based pharmacokinetic/pharmacodynamic modeling of the toxicologic interaction between carbon tetrachloride and kepone. Arch. Toxicol. 70:704-713.

Fisher, J., M. Lumpkin, J. Boyd, D. Mahle, J.V. Bruckner, and H.A. El-Masri. 2004. PBPK modeling of the metabolic interactions of carbon tetrachloride and tetrachloroethylene in B6C3F1 mice. Environ. Toxicol. Pharmacol. 16:93-105.

Gargas, M.L., R.J. Burgess, D.E. Voisard, G.H. Cason, and M.E. Andersen. 1989. Partition coefficients of low molecular-weight volatile chemicals in various liquids and tissues. Toxicol. Appl. Pharmacol. 98:87-99.

Ginsberg, G., D. Hattis, B. Sonawane, A. Russ, P. Banati, M. Kozlak, S. Smolenski, and R. Goble. 2002. Evaluation of child/adult differences from a database derived from the therapeutic drug literature. Toxicol. Sci. 66:185-200.

Gregory, G.A., E.I. Eger, and E.S. Munson. 1969. The relationship between age and halothane requirement in man. Anesthesiology 30:488-491.

Johnsrud, E.K., S.B. Koukouritaki, K. Divakaran, L.L. Brunengraber, R.N. Hines, and D.G. McCarver. 2003. Human hepatic CYPE1 expression during development. J. Pharmacol. Exp. Therap. 307:402-407.

Kim, H.J., J.V. Bruckner, C.E. Dallas, and J.M. Gallo. 1990. Effect of dosing vehicles on the pharmacokinetics or orally administered carbon tetrachloride in rats. Toxicol. Appl. Pharmacol. 102:50-60.

Manno, M., R. Ferrara, S. Cazzaro, P. Rigotti, and E. Ancona. 1992. Suicidal inactivation of human cytochrome P-450 by carbon tetrachloride and halothane in vitro. Pharmacol. Toxicol. 70:13-18.

McCarty, L.S., D. Mackay, A.D. Smith, G.W. Ozburn, and D.G. Dixon. 1991. Interpreting aquatic toxicity QSARs: The significance of toxicant body residues at the pharmacologic endpoint. Sci. Total Environ. 109/110:515-525.

Nagano, K., T. Nishizawa, S. Yamamoto, and T. Matsushimi. 1998. Inhalation carcinogenesis studies of six halogenated hydrocarbons in rats and mice. Pp. 741-746 in Advances in the Prevention of Occupational Respiratory Diseases, K. Chiyotani, Y. Hosoda, and Y. Aizawa, eds. Elsevier Science.

Rao, K.S., and R.O. Recknagel. 1968. Early onset of lipoperoxidation in rat liver after carbon tetrachloride administration. Exp. Mol. Pathol. 9:271-278.

Rao, K.S., and R.O. Recknagel. 1969. Early incorporation of carbon-labeled carbon tetrachloride into rat liver particulate lipids and proteins. Exp. Mol. Pathol. 10:219-228.

Smith, A.E., M.V. Evans, and D.M. Source. 1998. Statistical properties of fitted estimates of in vivo metabolic constants obtained from gas uptake data. I. Lipophilic and slowly metabolized VOCs. Inhal. Toxicol. 10:383-409.

Stevens, W.C. et al. 1975. Minimum alveolar concentrations (MAC) of isoflurane with and without nitrous oxide in patients of various ages. Anesthesiology 42:197-200.

Stewart, R.D., and Dodd, H.C. 1964. Absorption of carbon tetrachloride, trichloroethylene, tetrachloroethylene, methylene chloride, and 1,1,1-trichloroethane through the human skin. Amer. Ind. Hyg. Assoc. J. 25:439-446.

Thrall, K.D., M.E. Vucelick, R.A. Gies, R.C. Zangar, K.K. Weitz, T.S. Poet, D.L. Springer, D.M. Grant, and J.M. Benson. 2000. Comparative metabolism of carbon tetrachloride in mice, rats, and hamsters using gas uptake and PBPK modeling. J. Toxicol. Environ. Health 60:531-548.

COMMENTS ON SULFUR DIOXIDE

At its Feburary 21-23, 2005, meeting, the subcommittee reviewed the revised AEGL document on sulfur dioxide (SO_2). The document was presented by Peter Griem of Clariant, Germany. The subcommittee recommends a number of revisions.

General Comments

The NRC subcommittee recommends only relatively minor suggestions for improvement.

SO_2 is a substance that shares its delayed fatal response with e.g. paraquat and dioxane—that is, an initial virtual recovery followed by recurrence of serious life-threatening effects. There are many human data available; the subcommittee recommends that these be described more concisely.

Specific Comments

Page ii, line 34. Have any casualties or anaphylactic reactions been reported in asthmatics at low exposure?

Page 1, line 24. Add "co-exposure" before respirable particles at beginning of sentence.

Page 3, line 32. Because bronchiolar obstruction was still present 4 y after the accident, the airway obstruction is irreversible.

Page 4, line 1. What is "destructing bronchitis"?

Page 4, line 5. What happened to the follow-up after 4 y? Did patients escape medical supervision?

Page 5, line 6. In the London case, 1.3 ppm was the peak SO_2 concentration at which people died. This could lead to the conclusion that 1.3 ppm is an AEGL-3, whereas this document proposes 16-42 ppm for AEGL-3 and 0.75 ppm for AEGL-2 values. This discrepancy should be explained.

Page 5, lines 7-8. Change sentence to "The excess deaths were attributed to bronchitis or to other impairments of the respiratory tract."

Section 2.2.2. The database described here is fairly dated. The NAC should find more recent studies that have evaluated effects of SO_2. It is a given that epidemiological studies almost always have some confounders, but some of the studies have shown a stronger association to SO_2 exposure than other chemicals.

Page 7, line 5. The increase in sensitivity to SO_2 odor at the end of exposure is remarkable and the opposite for e.g. dioxane and hydrogen sulfide (H_2S) where the sensitivity of the olfactory system decreases after a certain time. Is this logical?

Page 7, line 12. Change "bronchiolar lavage" to "bronchoalveolar lavage" here and anywhere else it shows up.

Page 14, lines 40-41. If changes were shown to be statistically significant, how can it be difficult to ascertain the exact magnitude of the effect?

Page 19, line 6. Insert "ambient" before "air pollution."

Section 3.1. Are all these animal data really needed when there are so many human data available?

Section 4.4. This section does not seem to have a place in the document because the data described is not used for derivation of the AEGL values.

Section 4.5. Delete from text as explained below.

AEGL-1

The comment on page 27, line 32, that 0.25 ppm may be a threshold contradicts the comment on page 28, line 3, that effects were found at 0.25 ppm. In any case, it must be realized that effective concentrations in asthmatics are highly dependent upon the severity of the disease in the subjects being tested, the extent of medication use, etc. Thus, one study may show an effect at a certain concentration, and another study might show no effect merely because of differences in the subjects. Asthmatics are a highly variable group in terms of response to exposure to irritants, much more so than normal individuals exposed to the same atmospheres. Furthermore, most controlled clinical studies generally use subjects who are not the most severe. Based upon all this, the subcommittee concludes that the value for AEGL-1 of 0.25 ppm is too high and should be reduced to account for susceptibility differences in the most susceptible population, namely asthmatics. The subcommittee recommends a value of 0.2 ppm at the highest. The subcommittee recommends that the AEGL values should be held constant for various exposure durations.

Furthermore, the subcommittee recommends that the Comparative Indices table (Table 6) should be deleted from the document. This table could be misleading. For example, while an increase in SRaw of 200% may not be of concern in normals, it would surely be of concern in someone with preexisting respiratory disease. Thus, the comment on page 28, line 2, that a change of 134-139% is mild to moderate should be deleted from the text.

AEGL-2

The argument above for AEGL-1 applies here also. Changes in airway resistance of almost 600% is not necessarily of little consequence to an asthmatic.

Table II. The AEGL-3 is almost twice that of the Emergency Response Planning Guideline (ERPG-3). Thus, the latter seems to be more conservative. Some comment on this should be made. Similarly, the IDLH is more than twice the ERPG-3 value. In this case, the latter seems to be much more conservative.

Page 30. The AEGL-3 is extremely high in comparison to AEGL-2. Are there any examples of other substances that the subcommittee reviewed where such a high AEGL-3-to-AEGL-2 ratio exists? At first sight, the ERPG-3 seems to be more reasonable.

Conclusions of the Subcommittee

The AEGL-1 should be set at 0.2 ppm across the time scale.
The AEGL-2 should be 0.75 ppm throughout.
The AEGL-3 should remains as proposed *provided* a good justification for these values can be given.

COMMENTS ON CIS, TRANS 1,2-DICHLOROETHYLENE

At its February 21-23, 2005, meeting, the subcommittee reviewed the AEGL document on cis, trans 1,2-dichloroethylene. The document was presented by Cheryl Bast of Oak Ridge National Laboratory. The subcommittee recommends minor revisions to the document. A revised draft can be finalized if the recommended revisions are made appropriately.

Specific Comments

Page 24, lines 5-12. It should also be pointed out in this section that PBPK modeling can be very useful for time scaling in the derivation of AEGLs. Barton et al. (1995) published a model that was used to predict interactions between *trans*-1,2-dichloroethylene and other halocarbons, but it has not been validated for humans. $C^n \times t = k$ does a reasonable job for time scaling, from 4-8 h for AEGL-2 values but not for AEGL-3 values, as described below.

Page 25, lines 4-8. Derivation of AEGL-1: Data on narcosis and lethality were used for introducing a modifying factor for the derivation of the AEGL-1. Narcosis is clearly more severe than an AEGL-1 effect. It may therefore be recommended to justify this (for example, by the argument that the observation of a slight dizziness was taken as a possible mild narcotic effect at the 1,2-dichloroethylene concentration used as the starting point for the derivation of the AEGL-1).

Page 26, lines 10-37. It is not clear from the text in this section how the 10- and 60-min AEGL-2 values were derived. A short paragraph should be added that includes the experimental basis, departure point, and rationale for maintaining the AEGL at 1,000 ppm across these exposure periods.

As stated in the subcommittee's previous review of this chemical, Filser and Bolt (1979) predicted that *cis*- and *trans*-DCE would attain near-steady-state within 2 h in the blood of rats inhaling 100 ppm. This is characteristic of many VOCs. DCE is somewhat unique, however, in that it inhibits its own metabolism. This should result in some reduction in its systemic uptake. More importantly, the suicide inhibition could result in a continuing increase in blood concentrations (and CNS effects) at exposure periods longer than 6 h (that is, the basis for calculations). This logic supports the retention (as in the current draft of the document) of the 8-h AEGL-2 values of 450 and 230 ppm for *trans*- and *cis*-DCE, respectively, that were derived using $C^n \times t = k$.

Page 26, lines 19-21; page 28, lines 6-11 and 16-17; the Summary on page 5, lines 28-31; page 6, lines 8-11; the Appendix B tables page B-3, line 31ff; page B-5, line 32ff; page B-8, line 31ff; and page B-10, line 33ff. Derivation of AEGL-2 and AEGL-3: The argument given for the choice of the magnitude of the interspecies UF considers only pharmacodynamics: "because data suggest that the critical brain concentration of a halocarbon required to produce a given level of narcosis is relatively constant across species." Upon equivalent inhalation exposures, rats receive a substantially larger internal dose of VOCs than do humans. This is attributable to rats' higher respiratory rate and cardiac output. The influence of these is offset to a presumably quantifiable extent by the rats' more rapid metabolism of DCE. A comparative (approximate) quantification of these influences would be desirable. If the result would show that humans are at less risk than the rat (as this reviewer guesses), a further reduction of the interspecies UF would be warranted (e.g., from 3 to 1). If it should turn out impossible to estimate quantitatively the relative contributions of the essentially opposing control factors mentioned above, the pharmacokinetic contribution to the UF should remain at its full extent of 3 (or 4 to account for the generally greater impact of pharmacokinetic compared with pharmacodynamic differences; compare SOP manual). The intraspecies UF of 3 proposed by NAC appears justified. The subcommittee recommends using a well-justified modifying factor rather than trying to squeeze UFs if the final outcome does not stand the reasonability test.

Page 28, lines 17-22. Use of $C^n \times t = k$ does not do so well in scaling from 4 to 8 h for AEGL-3 values. The proposed 8-h AEGL-3 is half that of the 4-h value. The DCE blood and brain concentrations, as described above, will continue to increase with increasing duration of exposure. Exhalation of this very volatile chemical, however, will partially offset its systemic accumulation. Thus, it is very doubtful that DCE's blood concentration and CNS depressant effects will be twice as great at 8 as at 4 h (as indicated by the recommended 4- and 8-h AEGL-3 values).

Minor Points

Page 8, line 21. Perhaps change Budvari et al. (1989) to O'Neil et al. (2001). If Budvari should be retained, add to the References.

Page 10, line 22. The vapor concentrations should be expressed in ppm rather than ppm/m^3.

Page 10, line 32. Data in Table 7 are from a cat study by Lehmann and Schmidt-Kehl (1936) not the mouse study by Gradiski et al. (1978).

Page 12, lines 17-18. What is meant by "various researchers"? Did researchers other than Lehmann and Schmidt-Kehl conduct experiments that are described in this paragraph?

Page 17, line 4, and page 19, line 8. It's dichlorethene, not dichlorethane.

Page 21, line 28. Barton et al. 1995 not 1985.

Page 21, lines 38-42. Cite the appropriate reference in this paragraph.

Page 22, line 13. "other cytochrome P-450 substrates [not substances]"

Page 22, lines 27-28. Unit?

Page 23, lines 15-16. The ratios of AD_{95}: AD_{50} defines steepness of the dose response curve, not intraindividual variability. The latter may be estimated from the ratio of the highest to lowest maximal allowable concentration (MAC) to induce a given anaesthetic dose (AD).

COMMENTS ON MONOCHLOROACETIC ACID

At its February 21-23, 2005, meeting, the subcommittee reviewed the AEGL document on monochloroacetic acid. The document was presented by Peter Griem of Clariant, Germany. The subcommittee recommends minor revisions to the document. A revised draft can be finalized if the recommended revisions are made appropriately.

Overall Comment

The monochloroacetic acid document is greatly improved. With minor revisions, this AEGL TSD can be finalized.

Specific Comment

Line 1084. "Aanvaaarde" should read "Aanvaarde" (it's Dutch for accepted).

COMMENTS ON CARBON MONOXIDE

At its February 21-23, 2005, meeting, the subcommittee reviewed the AEGL document on carbon monoxide (CO). The document was presented by Peter Griem of Clariant, Germany. The subcommittee recommends minor revisions to the document. A revised draft can be finalized if the recommended revisions are made appropriately.

General Comments

The AEGL values and TSD for CO were approved by the subcommittee at its February 21-23, 2005, meeting. Only one minor change to the document was suggested. This involved expanding the support for why a carboxyhemoglobin (COHb) value of 40% was selected as the basis for the derivation of AEGL-3. Given that the range of anywhere from 34-56% COHb does not cause lethal effects (page 49, line 1822), why pick 40%? The discussion in the text supporting this decision (page 49, lines 1825-1927) is weak and could be strengthened by providing additional reasons for why 40% (as opposed to 45% or

50%, for example) was selected. Perhaps there are human data that support this decision in addition to the animal data cited in the text.

COMMENTS ON FLUORINE

At its February 21-23, 2005, meeting, the subcommittee reviewed the AEGL document on fluorine. The document was presented by Cheryl Bast of Oak Ridge National Laboratory. The document can be finalized.

COMMENTS ON METHANOL

At its February 21-23, 2005, meeting, the subcommittee reviewed the AEGL document on methanol. The document was presented by Peter Griem of Clariant, Germany. The subcommittee recommends the following revisions. The revised document will be reviewed at the subcommittee's future meeting.

Major Comments

An interspecies UF of 1 is used to derive AEGL-2 values even though the key study was a mouse study. The text states that an interspecies UF of 1 was selected because a sensitive species was used and because toxicokinetic differences between species were accounted for by using a PK model for calculating exposure concentrations (page 115, lines 3559-3563).

With regard to the use of sensitive species, it is not clear that the mouse is more sensitive than a human when considering developmental effects. The discussion of the NTP study panel that reviewed the data on developmental toxicity of methanol in rodents clearly states that the NTP panel concluded that there was "insufficient evidence to determine if the human fetus is more or less sensitive than the most sensitive rodent species (the mouse) to methanol teratogenesis" (page 51, lines 2050-2052). Since a developmental study was used to derive the AEGL-2 values, it seems inappropriate to use a UF of 1 in this case.

With regard to the use of PK modeling, this method does not completely eliminate all the uncertainty when extrapolating from animals to humans. PK modeling only takes into consideration a portion of the potential toxic impact of a substance. It does not address the pharmacodynamic component, which remains uncertain.

Unless there are data that clearly shows that animals (in this case mice) respond in the precise way that a human does to the same internal dose, then some uncertainty remains, and a UF greater than 1 needs to be used. The evidence presented supports moving away from a default of 10, but it is not convincing nor sufficient to use a UF of 1. An interspecies UF of 3 seems appropriate.

NOTE: There is no guidance in the SOP manual on how to incorporate PK modeling into the AEGL review process. This needs to be done as soon as possible because the use of these models is becoming more common.

Other Comments

AEGL-3 is based on a value set by the American Academy of Clinical Toxicology of the peak blood methanol concentration that indicates serious poisoning for which hemodialysis is recommended (page 78, lines 2228-2232). There is a general presumption that it is based on human data and general experience, but this could be made more explicit.

Specific Comments

Page 1, line 422. "tert.-butyl" is awkward. The NAC should either take the period out after "tert" or spell out tertiary.

Page 2, lines 454 and 470. WHO (1977) is listed as WHO (1997) in the References. Which is correct?

Page 2, lines 462-463. This sentence implies that bilateral blindness results from, or is directly related to, systemic metabolic acidosis. Tephy and his colleagues have clearly shown that optic nerve damage results from in situ (retinal) methanol metabolism.

Page 8, line 648. Typo: "significantly" should be "significant."

Page 8, lines 648-650. This sentence is confusing.

Page 9, line 728. How would experimental design limit one's ability to tolerate methanol inhalation?

Page 11, lines 800-805. As related in the subcommittee's initial review of this document, "A person who inhales ethanol vapor cannot attain blood levels sufficient to impair his/her ability to drive a motor vehicle. Methanol is less lipophilic and is therefore a less potent CNS depressant than ethanol." It is very unlikely that substantial CNS depression will result from the inhalation of methanol. Thus, it should be pointed out here that manifestations of poisoning would be limited to systemic acidosis and visual impairment.

Page 12, lines 812-819. Dermal exposure was obviously quite important in the methanol poisoning incident described by Humperdick (1941).

Page 19, line 1045, and page 10, lines 1062 and 1065. What are "responsive" stellate cells?

Page 26, line 1277. Should Rogers et al. (1997) be Rogers and Mole (1997)?

Page 28, line 1357. What is meant by the OECD guideline?

Page 30, lines 1420-1432. More information should be provided about the findings of Pollack and Brouwer (1996). For example, was the systemic clearance dose dependent, and how did fetal and maternal methanol concentrations compare?

Page 33, line 1512. Osterloh et al. (1996) is not included in the References.

Page 44, line 1790. This introductory sentence is incomplete. Something is missing from the sentence that ends with "between."

Page 46, line 1850. Clarify length of NIOSH (1980) and Frederick et al. (1984) study; it states only that it was conducted "during about 3 years."

Page 48, lines 1938-1947. The use of a UF of 3 does not appear to be warranted for the derivation of AEGL-1 values. No mucus membrane irritation or CNS effects were reported by human subjects exposed to 800 ppm for 8 h (Batterman et al. 1998). NIOSH (1980) and Frederick et al. (1984) described subjective symptoms reported by teachers and teachers' aides who were exposed to methanol vapors around duplication machines. These workers reported eye and throat irritation, headache, dizziness, etc. If the AEGL-1 values are based on mucus membrane irritation, the mean vapor concentration (1,060 ppm) should be utilized and the AEGL-1 "flat lined" across all exposure times. A UF is not employed for mild irritant effects. If the AEGL-1 values are based on CNS effects (that is, parent compound), a PBPK model should be used to derive AEGL-1 values for the different exposure times. Use of $C^n \times t = k$ (Haber's Rule) to extrapolate from longer to shorter exposures typically results in the underestimation of the risks of CNS depression or the overestimation of AEGL values for the shorter periods (Boyes et al. 2000; Bruckner et al. 2004). An interspecies UF is not needed for minimal CNS effects.

Page 48, lines 1946-1947. It is stated that "subpopulations with less than optimal folate status may be more susceptible to the health effects of methanol." This statement is not supported by several studies (cited in lines 1828-1831 on page 45) that show little effect of folate deficiency on blood methanol concentrations. CNS depression, of course, is directly dependent upon blood/brain methanol concentration.

Page 56, lines 2226-2228. It should be noted that exhalation is the second route of elimination of methanol, particularly at high exposure concentrations.

Page 68, line 2544. Typo: "live" should be "life."

Page 68, line 2547. Typo: Delete "a" before "two studies."

Page 68, line 2549. Typo: Remove "d" on derived.

References. Remove "cited in etc." from references. The original papers should be obtained, reviewed, and listed.

Additional References

Boyes, W.K., P.J. Bushnell, K.M. Crofton, M. Evans, and J.E. Simmons. 2000. Neurotoxic and pharmacokinetic responses to trichloroethylene as a function of exposure scenario. Environ. Health Perspec. 108(Suppl. 2):317-322.

Bruckner, J.V., D.A. Keys, and J.W. Fisher. 2004. The acute exposure guideline level (AEGL) program: Applications of physiologically based pharmacokinetic modeling. J. Toxicol. Environ. Health 67(Part A):621-734.

COMMENTS ON PHENOL

At its February 21-23, 2005, meeting, the subcommittee reviewed the AEGL document on phenol. The document was presented by Peter Griem of Clariant, Germany. The subcommittee recommends minor revisions. A revised draft can be finalized if the recommended revisions are made appropriately.

Specific Comments

Page 15-16. Mention that animals were exposed "whole body" in the following studies: Flickinger (1976) (line 736), Brondeau et al. (1990) (line 754), and Dalin and Kristofferson (1974) (line 795). The subcommittee suggests this to be consistent with the other inhalation studies in this section, where animals were exposed "nose only" or "whole body."

Page 15, line 736. Add in this sentence that rats were exposed to nominal concentrations of phenol aerosols. (It is elaborated on in the paragraph below, but the subcommittee believes that it helps to mention it in the first sentence.)